AA001145

MATERIALS RESEARCH SOCIETY
SYMPOSIUM PROCEEDINGS VOLUME **1564**

Single-Dopant Semiconductor Optoelectronics

April 1-5, 2013
San Francisco, California, USA

Printed from e-media with permission by:

Curran Associates, Inc.
57 Morehouse Lane
Red Hook, NY 12571
www.proceedings.com

ISBN: 978-1-63266-146-3

Some format issues inherent in the e-media version may also appear in this print version.

©Materials Research Society 2013

This reprint is produced with the permission of the Materials
Research Society and Cambridge University Press.

This publication is in copyright, subject to statutory exception and to the
provisions of relevant collective licensing agreements. No reproduction
of any part may take place without the written permission of Cambridge
University Press.

Cambridge University Press
Cambridge, New York, Melbourne, Madrid, Cape Town,
Singapore, São Paulo, Delhi, Tokyo, Mexico City

Cambridge University Press
32 Avenue of the Americas, New York, NY 10013-2473, USA
www.cambridge.org

Materials Research Society
506 Keystone Drive, Warrendale, PA 15086
www.mrs.org

CODEN: MRSPDH

ISBN: 978-1-63266-146-3

Cambridge University Press has no responsibility for the persistence or
accuracy of URLs for external or third-part Internet Web sites referred to
in this publication and does not guarantee that any content on such Web sites
is, or will remain, accurate or appropriate.

Additional copies of this publication are available from:

Curran Associates, Inc.
57 Morehouse Lane
Red Hook, NY 12571 USA
Phone: 845-758-0400
Fax: 845-758-2634
Email: curran@proceedings.com
Web: www.proceedings.com

TABLE OF CONTENTS

Simulation of a Single Dopant Nanowire Transistor.. 1
Asen Asenov, Vihar Georgiev

Theoretical Studies of Single Magnetic Impurities on the Surface of Semiconductors and Topological Insulators 13
M. R. Mahani, A. Pertsova, C. M. Canali, M. F. Islam, A. H. MacDonald

Author Index

Mater. Res. Soc. Symp. Proc. Vol. 1564 © 2013 Materials Research Society
DOI: 10.1557/opl.2013.1078

Simulation of a single dopant nanowire transistor

Asen Asenov[1,2] and Vihar Georgiev[1]

[1]Device Modeling Group, School of Engineering, University of Glasgow, G12 8LT, Glasgow, UK
[2]Gold Standard Simulations, Ltd., G12 8LT, Glasgow, UK

ABSTRACT

Nowadays the silicon technology is capable of delivering sub-10 nm devices where 'every atom counts'. Manipulation of atoms with high precision on such a scale, in principle, can lead to technological innovations, such as transistors with extremely short gate length, quantum computing components and optoelectronic devices. One possible strategy to create this next generation of devices is to precisely place individual discrete dopants (such as phosphorous atoms) in a nanoscale transistor.

In this paper, we report a systematic study of quantum transport simulation of an impact that precisely positioned dopants have on the performance of ultimately scaled gate-all-around silicon nanowire transistors (SNWT) designed for digital circuit applications. Due to strong inhomogeneity of self-consistent electrostatic potential, a *full* 3-D *real-space* Non Equilibrium Green's Function (NEGF) formalism is used. The simulations are carried out for an n-channel NWT with a 2.2 x 2.2 nm^2 cross-section and a 6 nm channel length, where locations of precisely arranged dopants in the source drain extensions and in the channel region have been varied. The individual dopants act as localized scatters and, hence, impact of electron transport is directly correlated to the position of the single dopants. As a result, a large variation in the ON-current and modest variation of the subthreshold slope are observed in the I_D-V_G characteristics when comparing devices with microscopically different discrete dopant configuration. Introducing of channel surface roughness in the Ch Sym 1 wire induces a threshold voltage shift and ON-current variation in the device due to scattering. The variations of the current-voltage characteristics are analyzed with reference to the behaviour of the transmission coefficients. Our calculations provide guidance for a future development of the next generation components with sub-10 nm dimensions for the semiconductor industry.

INTRODUCTION

Nanotechnology provides powerful tools and techniques to control individual atoms or molecules in atomic scale structures and devices. Manipulation of atoms with high precision on such a scale, in principle, can lead to technological innovations, such as transistors with extremely short gate length [1], quantum computing components [2] and optoelectronic devices [3].

One possible strategy to create this next generation of devices is to controllably introduce discrete atoms (dopants), such as phosphorus atoms, in a silicon crystal. Recent research, where a single phosphorous atom was embedded within epitaxial silicon environment, analysis a possibility of creating a single-atom transistor [4]. Although such an idea looks spectacularly attractive, the side gate architecture used in Ref 4 is not really fit for the purpose of digital applications. In order to be able to reliably switch on/off such nanometer scale transistors the preferable transistor architecture needs to be based on a gate wrapped around concept [5]. Indeed such all-gate-around silicon nanowire transistors (SNWT) have not only been experimentally demonstrated but have also been extensively studied revealing the impact of a single channel dopant on the device performance [6]. In this paper we answer the following important question 'from the point of view of digital application, is it more beneficial to have a single dopant in the channel of a SNWT or not to have a dopant at all?"

In order to answer the above question, we have carried out calculations using an n-type SNWT transistor with 6 nm channel length and with a 2.2 x 2.2 nm^2 cross-section, which is illustrated in Fig. 1.

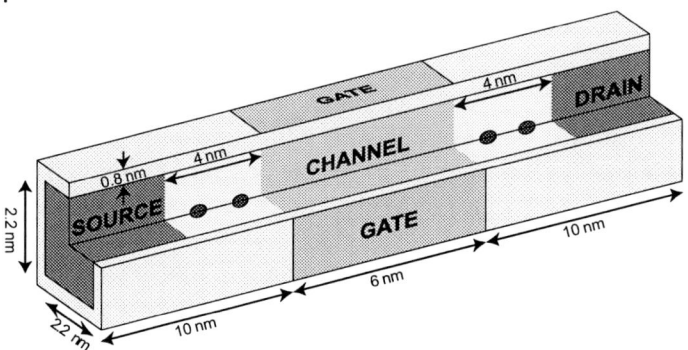

Fig. 1 Schematic view of the gate-all-around nanowires, showing dimensions of the transistor.

Many researchers expect that transistors with such dimensions will mark the limit of CMOS scaling. The important features of such transistors are the unavoidable 'spacers' between the source/drain contacts and the gate, which have to be doped (white rectangular regions in Fig. 1). Based on the volume of the spacer (access) regions and the solid solubility of phosphorous in Si, for our SNWT dimensions, there are on average 2 dopants on each side of the gate. It has been demonstrated already that the randomness of the dopants in the access region can introduce significant variably in the characteristics of the SNWT [7]-[9]. Here, however, we will assume that future technology innovations will allow us to align the dopants in the most favorable (from the device performance point of view) position marked 'Ch Sym' in Fig. 2.

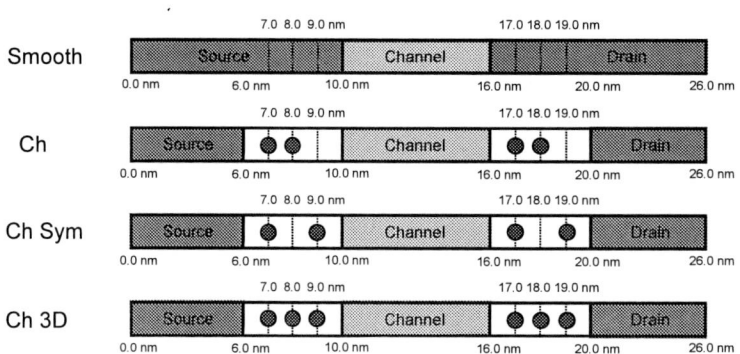

Fig. 2 Schematic view of the Smooth, Ch, Ch Sym and Ch 3D devices used in this work.

This Ch Sym transistor with undoped channel will be our reference point, which will allow us to compare performance of analogous transistors, such as Ch and Ch 3D presented in Fig. 2. The device marked as 'Smooth' is introduced in order to provide comparison with our previously published works [10], [11].

This work begins with a brief description of the self-consistent NEGF/Poisson algorithm in the theory section. The following part (discussion) presents results of the simulations and it is divided into three subsections. The first subsection reveals the impact of precisely placed dopants that are in the access region (see Fig. 2). It also compares these dopants with random doping. In the second part of the discussion section we investigate the effect of a single channel dopant and its position along the channel on the device performance. The third part of the same section reveals the combined effect of channel surface roughness and discrete dopant position in the Ch Sym 1 SNWT on the current-voltage characteristics. In the conclusion section we summarize the main points of this work.

THEORY

Simulations are carried out using the quantum transport module of the GSS TCAD simulator GARAND [12]. Accurate modeling of such small transistors with channel lengths of sub-10 nm, particularly in the presence of discrete dopants, requires a full 3-D Non-Equilibrium Green's Function (NEGF) quantum transport treatment due to strong quantum confinement and tunneling [13-15]. In our work we neglect all sources of incoherent scattering, such as phonon interaction. Based on experimental evidence and previous research we believe that the inclusion of phonon scattering would have little impact on our results [16]. Moreover, in the highly doped source and drain regions the scattering is dominated by impurity scattering and phonons play a minor role. The random rough interface between Si and SiO_2 is introduced using the approach described in Ref. [8]. Any roughness between the oxide and the gate material is ignored.

The Hamiltonian used in the discretization of the NEGF equations is the effective-mass Hamiltonian that folds the full crystal interaction into the electron effective masses. The effective masses of the valleys are extracted from tight binding calculations that capture the dependence of the electron band structure on the nanowire cross-section. Due to the small cross-section of the SNWT, only four of the six valleys of the silicon conduction band were included. The two valleys that were neglected have transversal masses ($0.3m_e$) in the direction perpendicular to the wire axes resulting in a larger ground-state-energy shift. As a result, the electron population in these valleys is negligible if compared to the other four valleys for the simulated nanowire orientation, diameter, temperature, and bias conditions. The correlation matrix, $G^<$, was calculated using a recursive algorithm. Boundary conditions of the Green's function equations at the contacts, which are given throughout the contact self-energies, were defined by using the algorithm described in reference [14]. From the correlation matrix, the electron and current densities are calculated by the following equations:

$$n(E,x) = iG^<(E,x,x) \qquad (1)$$

$$J(E,x) = -i\frac{e\hbar}{2m}(\nabla - \nabla')G^<(E,x,x)\Big|_{x=x'} \qquad (2)$$

where $n(E,x)$ and $J(E,x)$ are the electron and current densities, respectively. The electron density is used to calculate, self-consistently, the electrostatic potential through the Poisson's equation. The obtained solutions of the NEGF and Poisson equations are iterated until density and current converge.

DISCUSSION

In the first two parts of the discussion section we will briefly summarise the results that have been published in our recent work [11]. This information is important for the consistency and better understanding of this paper. These discussions are followed by previously unpublished results of on the simulations of a single dopant SNWT including channel surface roughness.

Impact of precise dopant placement in the source and drain regions on device performance

As a first step in our study we carried out simulations on fifty devices with random dopants in the source and drain region reflecting the traditional way of introducing doping by using ion implementation and annealing. The current-voltage characteristics for all of these fifty microscopically different SNWTs (with various impurity configurations) are presented in Fig. 3.

The results reveal that the presence of random dopants in the access regions leads to fluctuation of the threshold voltage and the OFF- and ON-current. Moreover, the OFF-current in most of the devices is lower if compared to the smooth transistor. Additionally, the random dopant configurations in the access regions also result in threshold voltage and subthreshold slope variations but these are relatively small. Furthermore, the average magnitude of the ON-current drops by 43% (blue line in Fig. 4) in comparison to the uniformly doped purely ballistic devices (smooth). Hence, it can be concluded that the presence of random dopants in the access regions has significant influence on the device characteristics.

In an attempt to reduce the ON-current variability and to improve the transistor performance we have created nanowire transistors with precisely placed discrete dopants in the source/drain (S/D) regions. The transistors have the dopants positioned in the middle of the nanowire and aligned along the channel – Ch, Ch Sym and Ch 3D, which are illustrated at the bottom of Fig. 2. Two of the devices presented in Fig. 2, Ch and Ch Sym, have four precisely placed dopants, which correspond closely to the average number of dopants in the access regions. The only difference between those two SNWTs is the position of two phosphorous atoms along the channel. From our previous work and the simulations presented earlier in this section, it is clear that the positions of the impurities can dramatically impact the ON-current without significantly affecting the threshold voltage and the electrostatic integrity (subthreshold slope). For example, the symmetrically placed dopants, in the case of Ch Sym, result in high ON-current, which is above the average value obtained from the random dopant simulations. In the Ch Sym transistor, the ON-current is 71% of the maximum possible ballistic value (smooth device). It is important to emphasise that in the case of random dopants the average ballisticity in this 6 nm channel length transistor is only 57%. Additionally, arranging the dopants asymmetrically in the S/D, in the Ch configuration results in a dramatic reduction in the ON-current. In such a transistor the 'drive' current has the lowest value among all transistors and the ballisticity coefficient drops to 33%.

Further improvement in the transistor performance can be achieved by increasing the number of precisely placed dopants in the access regions – Ch 3D. The 'highly doped structure' shows remarkably high level of balisticity, 81%. This is a 10% increase in the current if compared to the Ch Sym case.

In an attempt to further improve the behaviour of the devices with discrete impurities we introduced a single phosphorous atom to the channel

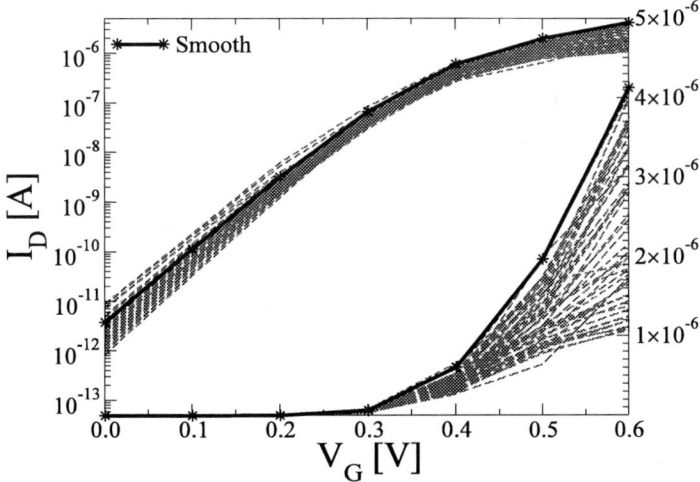

Fig. 3 I_D-V_G characteristics of 50 nanowire trasnsistors with discrete random dopants in the S/D region, a channel length of 6 nm, and a cross-section of 2.2 x 2.2 nm^2, V_D = 50 mV.

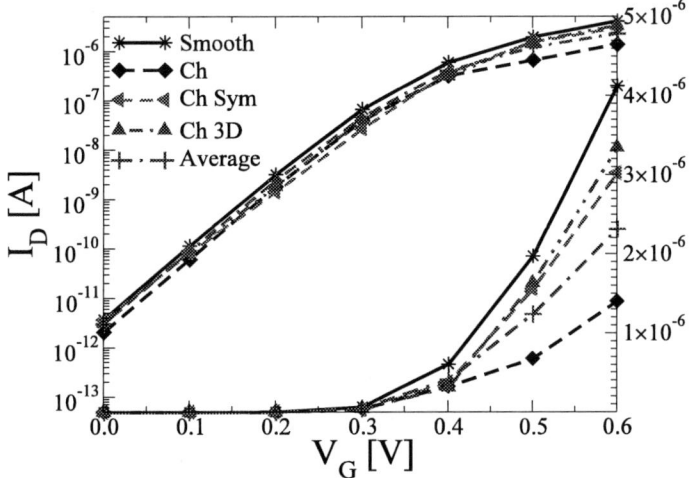

Fig. 4 I_D-V_G characteristics of nanowire trasnsistors with discrete precisely placed dopants in the S/D region for structures presented in Fig. 2. The blue line is the average value of the current obtained from Fig. 3.

Precisely placed single-channel dopant

There have been speculations that a single-donor dopant in the channel can increase the ON-current by creating a 'sombrero' potential [8] and a resonant current path between the source and the drain. In this section, we investigate the effect of a single phosphorous atom in the channel region on the performance of the SNWTs. Four test devices are considered (Ch Sym 1, Ch Sym 2, Ch Sym 3 and Ch Sym 4) and they are schematically illustrated in Fig. 5. All four wires have symmetrically placed dopants in the S/D and one dopant in the channel.

An analysis of this figure reveals three important effects that the additional channel dopant has over the device performance.

Firstly, there is a threshold voltage reduction associated with the lowering of the potential barrier between the source and the drain by the dopant atom. The effect is more dramatic when the dopant is placed at the maximum of the barrier (Ch Sym 1) and almost negligible when the dopant is close to the source extension (Ch Sym 4). This is not a significant problem for the modern CMOS technology because the threshold voltage can be adjusted by gate work function engineering.

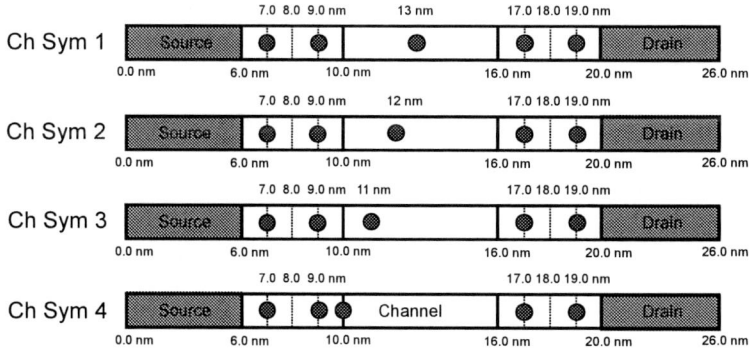

Fig. 5 Schematic view of single dopant transistors presented in this paper.

The current-voltage characteristics of simulated structures with a channel dopant, for the smooth and for Ch Sym transistors, are presented in Fig. 8.

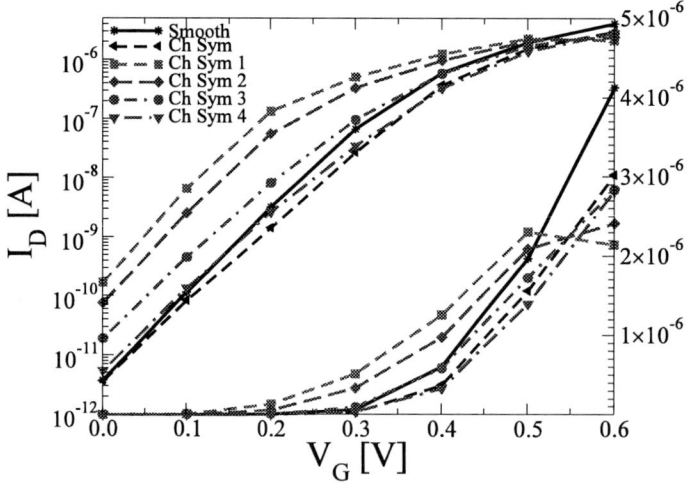

Fig. 6 I_D-V_G characteristics of nanowire trasnsistors a with single channel dopant presented in Fig. 5. The smooth and Ch Sym devices are also shown for comparison.

Secondly, there is a deterioration in the subthreshold slope, which increases when the dopant is moved from the source towards the center of the channel. This is indeed a bad news for digital electronic applications because in a billion transistor count chips the deterioration of the subthreshold slope can increase the leakage current even if the threshold voltage is adjusted to be the same as in the Ch Sym case.

Finally, the channel dopant increases the Coulomb scattering in the channel region, reducing the ON-current, which slows down the corresponding digital circuits. Despite the lower threshold voltage, which results in high gate voltage overdrive, all transistors with dopant in the channel, have lower 'drive' current at the anticipated supply voltage of 0.6V.

Clearly the transistor with a dopant in the middle of the channel is the worst device in relation to all three effects described above. It has the largest threshold voltage shift, the flattest subthreshold slope and the lowest 'drive' current.

The variation of the current-voltage characteristics can be understood with reference to the behavior of the transmission coefficient. Fig. 7 shows the transmission spectra for the Smooth, Ch Sym, and all single-dopant transistors at low $V_G=0.0V$ and high $V_G=0.5V$ gate voltages.

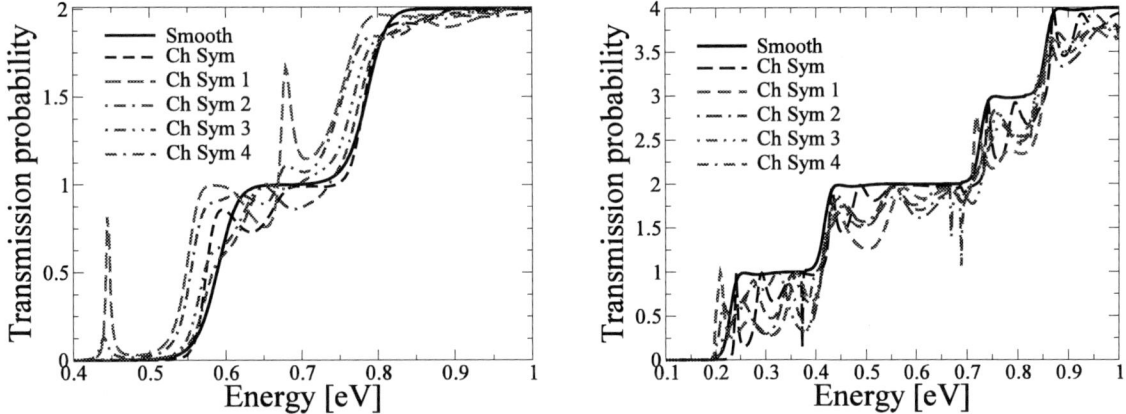

Fig. 7 Transmission prabability at $V_G=0.0V$ (left hand side) and $V_G=0.5V$ (right hand side) for all single-dopant transistors, Smooth and Ch Sym devices.

At $V_G=0.0V$, the discrete dopants lead to resonance peaks which are very well-pronounced in the Ch Sym 1 device – peaks around 0.45 and 0.7 eV. The transmissions spectra for the transistor rise first and, hence, the device has the highest leakage current at low gate biases. At $V_G=0.5V$, the curves are more jagged and translated to lower energies, due to increasing of scattering inside of the channel. For all transistors with discrete dopants the shape of the self-consistent potential around the impurities produces so-called Fano-type resonances, which appear as asymmetric dips and peaks in the transmission. The width, position and the number of the resonant peaks and antiresonant dips depends on the position of the impurity in the SNWT. Note that the transmission for the smooth device has a stair-type characteristic as expected. Therefore, the irregular shape and peaks/dips in the transmission for the devices with discrete impurities can be interpreted as a fraction of carriers being reflected back. This feature is not visible at lower gate bias because of the dominant reflection of the carriers by the channel barrier potential.

The reduction of the channel barrier potential (electrostatic potential) with increasing of the gate voltage is clearly visible in Fig. 8. The electrostatic potential along the channel is plotted in a white dashed line, which is aligned to the first sub-band energy at the source. This gives an opportunity to compare the position of the quasi-bound states relative to the inverted sombrero shape of the potential. Moreover, the resonances are clearly shown in the distribution of density of state (DOS) along the channel as presented in Fig. 8 for the Ch Sym 1 case. The DOS is higher around the discrete dopants and the transmission function (plotted on the left-hand side of Fig. 8 with white solid line) has peaks corresponding to the channel dopant energy levels. Applying high positive gate bias, such as 0.5V, leads to lowering of the potential barrier in the channel moving of the transmission spectra toward lower energies. For example, the first transmission peak moved almost 0.2 eV from 0.45 eV at $V_G = 0.0$ V to 0.25 eV at $V_G = 0.5$ V.

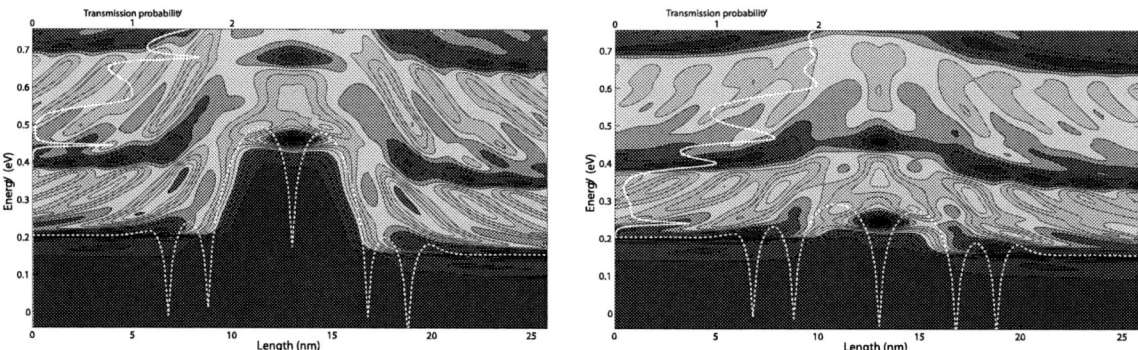

Fig. 8 Transmission probability (solid white line) and density of states along the wire for the Ch Sym 1 case for a gate voltages $V_G = 0.0$ V (left) and $V_G = 0.5$ V (right). The horizontal dashed white line is the electrostatic potential and it has been aligned to the first subband energy at the source.

Ch 1 Sym transistor with introduced channel surface roughness

In order to understand the unavoidable variability the transistor with the worst behavior from the previous section (Ch Sym 1) we introduced another source of statistical variability – *surface roughness*. In reality, *interface roughness* and *discrete dopants* will simultaneously be present in SNWT, and it is important to investigate the combined effect of the two sources of variability. Fig. 9 shows a schematic picture of a Ch 1 Sym SNTW with introduced interface roughness in the channel region.

To study the statistical effect associated with the *interface roughness*, we simulated and analyzed the I_D-V_G characteristics of 30 nanowire transistors with different randomly generated *interface roughness pattern*. The collection of individual I_D-V_G curves of those 30 devices is shown in Fig. 10. For comparison the smooth and Ch Sym 1 wire are also presented in Fig. 10.

An analysis of this figure reveals important features in the current-voltage characteristics related to the combined effect of *surface roughness and single channel dopant.*

Firstly, all devices with *surface roughness* have lower OFF-current in comparison to the Ch Sym 1 transistor. Hence, introducing of a variation of the channel thickness with combination of the presence of a single dopant in the middle of the channel in principle could lead to decreasing of the leakage current. However, the Smooth device still has the lowest OFF-current relative to all single atom transistors.

Secondly, the ON-current variation is significant and most devices with rough interface have lower 'drive' current in comparison to the Ch Sym 1. However, it is possible to have a rough device with higher ON-current than the Ch Sym 1 case but probability for this to happen is low. Only five from all thirty transistors have higher ON-current at V_G=0.6 V if compared to the Ch Sym 1.

Lastly, we observe large spread in threshold voltage, which, for the 30 devices simulated here, spans around 120 mV. For the case with discrete dopants only (Fig. 1), the threshold voltage span is smaller - around 80 mV. Additionally, the combined effect of surface roughness and channel dopant shows changes in the subthreshold slope. Again, this effect is more pronounced in the 'rough' transistors in comparison to the nanowires than only have random discrete dopants.

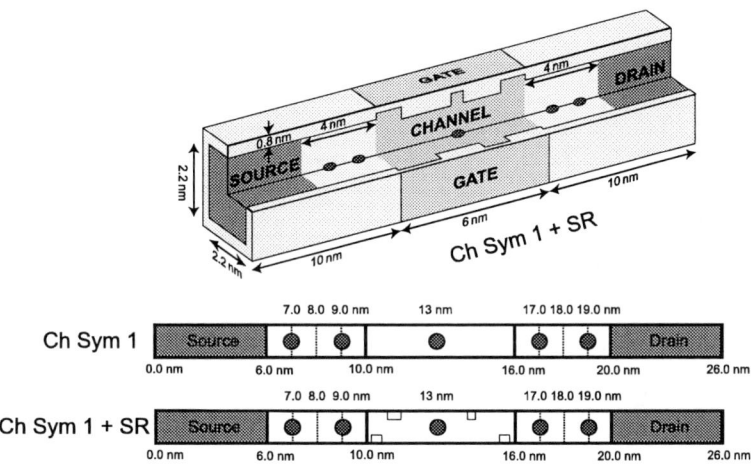

Fig. 9 Schematic view of the Ch Sym 1 transistor with *surface roughness* in the channel – Ch Sym 1 + SR.

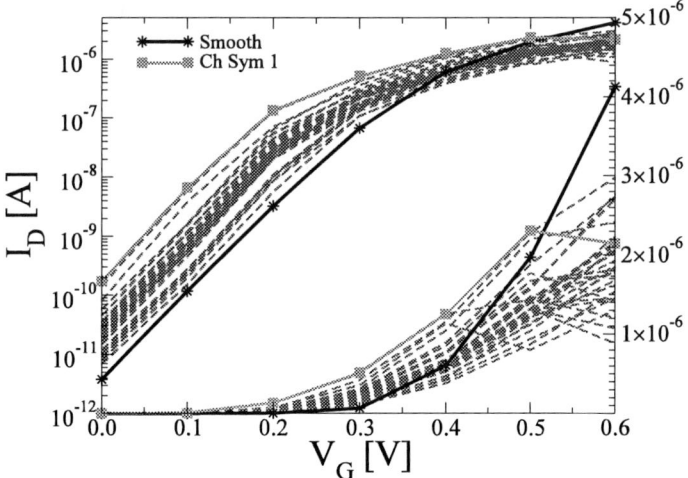

Fig. 10 I_D-V_G characteristics of the Ch Sym 1 + SR nanowire trasnsistors with various *surface roughness* in the channel. The smooth and Ch Sym 1 devices are also shown for comparison.

A more detailed explanation of the current-voltage characteristics cannot be further pursued without reference to the behavior of the transmission coefficient. We focused our analysis on three transistors that produce the lowest, median and highest current at 0.5 V gate bias. The transmission functions of those transistors at low ($V_G = 0.0$ V) and higher ($V_G = 0.5$ V) gate voltages as a function of energy are shown in Fig. 11.

At low gave bias of $V_G = 0.0$ V, introducing of *surface roughness* in a device with a single channel dopant is sufficient to rise the ground state of the transversal wave function, resulting in a step increase in transmission occurring at higher energies. For example, the first transmission peak for all SNWT transistors with *surface roughness* is shifted by around 0.05 eV from 0.45 eV for the Ch Sym 1 wire to 0.50 eV for all other devices. This is reflected in the I_D-V_G characteristics where the current for all 'rough' devices is lower in comparison to the Ch Sym 1 transistor at 0.0 V gate voltage.

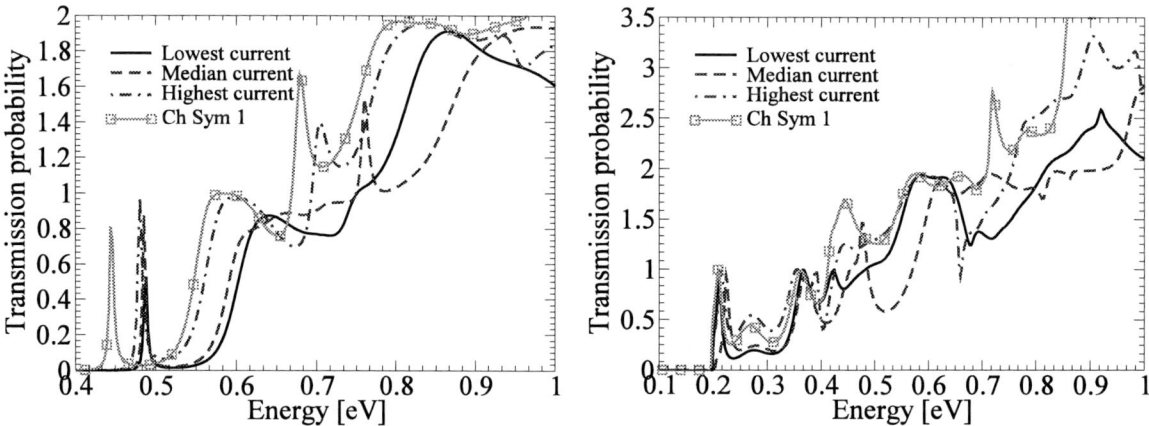

Fig. 11 Transmission prabability at V_G=0.0V (left) and V_G=0.5V (right) for the Ch Sym 1 and three *surface rougness* devices.

A similar analysis is also performed at higher gate voltages ($V_G = 0.5$ V) – Fig. 11. Similarly to Fig. 7 at V_G=0.5V, the curves are more jagged and translated to lower energies relative to the same spectra at V_G=0.0V, due to increasing of scattering inside of the channel. The transmission coefficients indicate the presence of Breit-Wigner- and Fano-type resonances. The width, position and the number of the resonant peaks and antiresonant dips depends on the profile of the surface roughness in combination with the discrete channel dopant in the SNWT. Similarly to the results in the previous subsection, those features are less pronounced at lower gate bias because of the dominant reflection of the carriers by the channel barrier potential.

Detailed information about the shape and the hight of the potential barrier for the Ch Sym 1 transistor *with* and *without surface roughness* can be obtained from Fig. 12. Fig. 12 presents the density of states (DOS), transmission spectra and the electrostatic potential for the Ch Sym 1 transistors at $V_G = 0.0$ V gate voltage *with* (right hand side) and *without* (left hand side) surface roughness. The electrostatic potential along the channel is plotted in a white dashed line, which is aligned to the first sub-band energy at the source. This gives an opportunity to compare the position of the quasi-bound states relative to the inverted sombrero shape of the potential. The transmission spectrum is plotted on the right hand side of each figure with a white solid line. The density of state at each case is presented by colorful 2D contour.

Introducing of the channel *surface roughness* leads to partial increasing and perturbing of the symmetric profile of the potential barrier in the channel. This is presented by the white dashed curved in Fig. 12, which is more jagged and perturbed on the right hand side in comparison to the left part of the same figure. The influence of surface roughness is even more clearly visible in the distribution of the density of states (DOS). Similar to Fig. 8, the DOS is higher around the discrete dopants and the transmission function has peaks around the energy levels of the discrete phosphorus atoms. Surface roughness causes a move of the energy states of the single channel dopant to higher energies and, as a result, the whole transmission function of the Ch Sym 1 with *surface roughness* transistor is shifted to higher energy. These effects are clearly visible for the first transmission peak, which moves by almost 0.03 eV from 0.45 eV for Ch Sym 1 to 0.49 eV for Ch Sym 1 with surface roughness. The second transmission peak at around 0.7 eV for the Ch Sym 1 structure is absent in the 'rough' transistor in the same energy scale. Moreover, the transmission spectrum for the 'rough' device has shorter peaks and significantly different profile in comparison to the 'rough free' Ch Sym 1 nanowire.

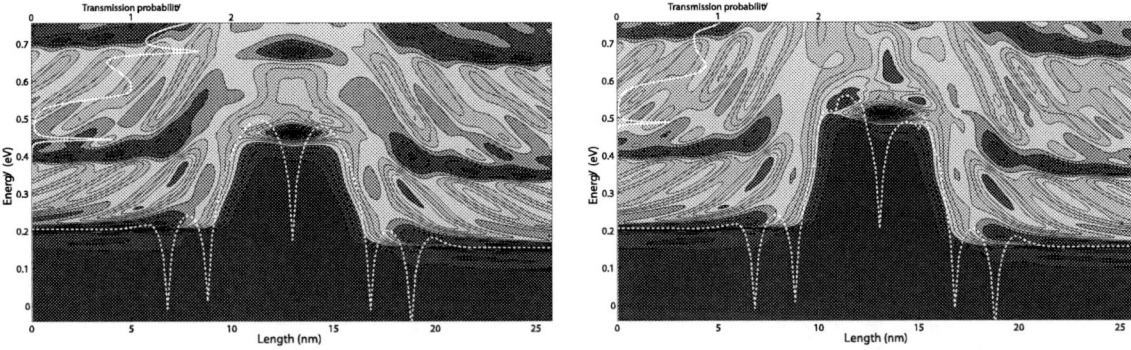

Fig. 12 Transmission probability (solid white line) and density of states along the wire for the Ch Sym 1 case *without* (left hand side) and *with surface roughness* (right hand side) with the lowest ON-current for a gate voltages $V_G = 0.0$ V. The horizontal dashed white line is the electrostatic potential and it has been aligned to the first sub-band energy at the source.

CONCLUSIONS

We performed quantum transport simulations of the impact of discrete phosphorous atoms and surface roughness on the operation of an all-gate-around silicon nanowire transistor (SNWT) in order to assess its potential application in digital circuits and systems.

The first part of the discussion section investigated the impact of the specific spatial configuration and of the number of discrete dopants in the transistor over device performance. The transistors that include only the S/D dopants show almost negligible difference in the subthreshold slope and the threshold voltage. On the contrary, we observed a general decrease and large variations in the 'drive' current associated with the position of the discrete dopants due to backscattering.

Furthermore, in the second part of the discussion section we introduced an individual phosphorous atom in the channel region of the transistor. This leads to an even stronger deterioration in the performance of the nanowire transistors in comparison to the undoped channel wires, such as Ch Sym 1.

Finally, in the last section of the discussion part, we included surface roughness in the Ch Sym 1 device in order to obtain realistic treatment of the devices performance.

All calculations presented in this work show the following common trends. Firstly, the ON-current that is critical for the circuit performance is worse for all transistors with discrete dopants and surface roughness considered in our study in comparison to the uniform doped devices. Secondly, the presence of backscattering and resonances, including very localized zero-transmission resonances deriving from unscreened Coulomb wells and surface roughness, is a common feature in the transmission functions. Thirdly, the OFF-current for all single atom devices with surface roughness is higher in comparison to the Smooth wire.

This paper shows the importance of quantum effects on SNWT not only in terms of the threshold voltage shift and ON- and OFF-current but also in terms of transport in the presence of surface roughness and ionized impurity scattering. Our results reveal a strong link and correlation between the undelaying microscopic structure of SNWT and the devices performance, which plagues devices with small channel length.

REFERENCES

1. D. J. Frank, R. H. Dennard, E. J. Nowak, P. M. Solomon, Y. Taur and H-S. P. Wong, Proc. IEEE 89, 259 (2001)
2. J. J. Morton, D. R. McCamey, M. A. Eriksson and S. A. Lyon, Nature 479, 345 (2011)
3. P. M. Koenraad and M.E. Flatte, Nature Mater. 10, 91 (2011)
4. M. Fuechsle, J. A. Miwa, S. Mahapatra, H. Ryu, S. Lee, O. Warschkow, L. C. L. Hollenberg, G. Klimeck and M. Y. Simmons, Nature Nanotech. 7, 242 (2012)
5. N. Singh, A. Agawal, L. K. Lakshmi, T-Y. Y. Liow, R. Yang, S. C. Rustagi, C. H. Tung, R. Kumar, G. Q. Quang, N. Balasubramanian, D. L. L. Kwong, IEEE Elec. Dev. Lett. 27, 383 (2006)
6. B. C. Johnson, G. C. Tettamanzi, A. D. C. Alves, S. Thompson, C. Yang, J. Verduijn, J. A. Mol, M. Vinet, M. Sanquer, S. Rogge and D. N. Jamieson, Appl. Phys. Lett. 96, 264102 (2010)
7. A. Martinez, N. Seoane, A. R. Brown, J. R. B. Barker and A. Asenov, IEEE Trans. Nanotechnol. 8, 603 (2009)
8. A. Martinez, N. Seoane, A. R. Brown, J. R. B. Barker and A. Asenov, IEEE Trans, Elec. Dev. 57, 1626 (2010)
9. N. Seoane, A. Martinez, A. R. Brown, J. R. B. Barker and A. Asenov, IEEE Trans, Elec. Dev. 56, 1388 (2009)
10. A. Martinez, M. Aldegunde, N. Seoane, A. R. Brown, J. R. B. Barker and A. Asenov, IEEE Elect. Dev. 58, 2209 (2011)
11. V. P. Georgiev, E. A. Towie, A. Asenov, IEEE Trans. Elec. Dev. 60, 965 (2013)
12. http://www.goldstandardsimulations.com/products/garand/
13. R. Landauer, Philos. Mag. 21, 863 (1970)
14. M. Buttiker, IBM J. Res. Dev. 32, 317 (1981)
15. Y. Imry and R. Landauer, Rev. Mod. Phys. 71, 306 (1999)
16. K. H. Hwan, K. H. Yeo, Y. Young, S.-D. Suk, M. Li, H. Jeong-Min, S.Min-Sang, D. W. Kim, D. G. Park, B. H. Hong, Y. C. Jung and S. W. Hwang, Appl. Phys. Lett. 92, 052 102 (2008)

Mater. Res. Soc. Symp. Proc. Vol. 1564 © 2013 Materials Research Society
DOI: 10.1557/opl.2013.1088

Theoretical studies of single magnetic impurities on the surface of semiconductors and topological insulators

M. R. Mahani[1] A. Pertsova[1], C.M. Canali[1], M. F. Islam[1] and A.H. MacDonald[2]
[1]Department of Physics and Electrical Engineering, Linnaeus University, Norra vägen 49, 391 82, Kalmar, Sweden.
[2]Department of Physics, University of Texas at Austin, U.S.A.

ABSTRACT

We present results of theoretical studies of transition metal dopants in GaAs, based on microscopic tight-binding model and *ab-initio* calculations. We focus in particular on how the vicinity of surface affects the properties of the hole-acceptor state, its magnetic anisotropy and its magnetic coupling to the magnetic dopant. In agreement with STM experiments, Mn substitutional dopants on the (110) GaAs surface give rise to a deep acceptor state, whose wavefunction is localized around the Mn center. We discuss a refinement of the theory that introduces explicitly the d-levels for the TM dopant. The explicit inclusion of d-levels is particularly important for addressing recent STM experiments on substitutional Fe in GaAs. In the second part of the paper we discuss an analogous investigation of single dopants in Bi_2Se_3 three-dimensional topological insulators, focusing in particular on how substitutional impurities positioned on the surface affect the electronic structure in the gap. We present explicit results for Bi_{Se} antisite defects and compare with STM experiments.

INTRODUCTION

The study of the spin of individual transition-metal dopants in semiconductor hosts and novel quantum materials, such as topological insulators, is an emergent field in nanoscience known as magnetic solotronics [1]. The capability of controlling, monitoring and manipulating single-spins in these materials is opening exciting prospects for novel spintronics [2, 3] and quantum computation devices at the atomic scale [4].

Among the most important experimental methods, advances in different STM-based techniques permit the investigation of substitutional dopants placed near the surface of a host material with unprecedented accuracy and degree of detail [5, 6]. It is therefore possible for theory to investigate these complex systems and elaborate descriptions which can be compared with experiments under highly controlled conditions.

In this paper we consider two examples of this theoretical effort, based on a multiband tight-binding model supplemented by *ab-initio* methods. In the first case, we examine magnetic ions in GaAs. When substituted for Ga atoms, the magnetic dopants act as spin center strongly coupled to the associated itinerant hole-acceptor states. We investigate in particular the nature of the midgap electronic states for the case of Mn and Fe on the (110) surface. The second example deals with Bi_2Se_3 which is one of the most studied three-dimensional topological insulators [7-10]. For this case we examine individual bismuth dopant substituting selenium atoms in the surface layer. We show that our tight-binding model is able to capture the salient features of the local density of states (LDOS) around the impurity as observed in STM experiments.

MAGNETIC IMPURITIES IN GaAs

Transition-metal (TM) impurities in III-V semiconductors act as spin centers, which interact strongly with the bound hole-acceptor carriers that they introduce in the host material. Mn and Fe dopants in GaAs are an important example. The spin of the magnetic ion couples antiferromagnetically with the spin and orbital moment associated with the valence hole from the GaAs host. The combined TM-magnetic core and acceptor spin and orbital moment form a novel nanomagnet, which we refer to an *acceptor magnet*. Because of the strong spin-orbit interaction, the properties of this magnetic entity can be effectively controlled electrically and addressed optically, via the manipulation of the hole wavefunction. A theoretical framework based on a microscopic tight-binding model is particularly convenient to investigate these properties.

Mn dopants in GaAs surfaces

In this section we present results of a theoretical model aimed at describing the electronic and magnetic properties of individual Mn dopants placed at or in the vicinity of the (110) surface of GaAs. We model substitutional Mn impurities in GaAs using a multi-band tight-binding (TB) Hamiltonian [11, 12]

$$H = H_{GaAs} + H_{Mn} + H_{Mn-As} + H_{SOI} \tag{1}$$

The first term is a Slater-Koster sp^3 TB model that reproduces the band structure of bulk GaAs. The parameters of the model are appropriately rescaled when needed to account for the buckling of the (110) surface. H_{Mn} describes the Mn 3p, 3d and 4s orbitals. The salient feature is the energy splitting between the spin-up and spin-down (or equivalently majority and minority with respect to a given quantization axis) d-levels, which we take from density functional theory (DFT) calculations [13]. The third term H_{Mn-As} describes the hybridization between the Mn orbitals and the orbitals of its nearest-neighbor As atoms. In the present study we assume that these parameters are equal to the parameters describing the hybridization between Ga and As atoms, including the important d-p hybridization. These values are again extracted from DFT calculations for GaAs [14]. Finally, the last term is a one-particle Hamiltonian describing the crucial spin-orbit interaction, which we take atomic in character.

The Hamiltonian in Eq. (1) differs in an important way from similar multi-band tight-binding model recently used to describe Mn impurities in GaAs [11, 12]. In Refs. [11, 12] the effective antiferromagnetic exchange coupling between the Mn spin and the nearest-neighbor As p-orbitals was described by an Heisenberg coupling between the quantum spins of the p-levels and a classical vector, representing the Mn. As such, that model did not contain the d-levels of the Mn. In the present model, the effective exchange interaction results from the hybridization of the Mn (exchange split) d-levels and As p-levels. This is an important improvement in order to describe TM impurities other than Mn.

The electronic structure of GaAs with a single substitutional Mn is obtained by performing supercell calculations with a cubic cluster, typically of 3200 atoms and periodic boundary conditions in either 2 or 3 dimensions, depending on whether we are studying the (110) surface or bulk. In order to remove artificial dangling bond states appearing in the gap, we include relaxation of surface layer positions. To assess finite-size effects on our results we have also employed Lanczos methods in much larger systems (70,000 atoms), which allow us to investigate a small energy window around the acceptor level.

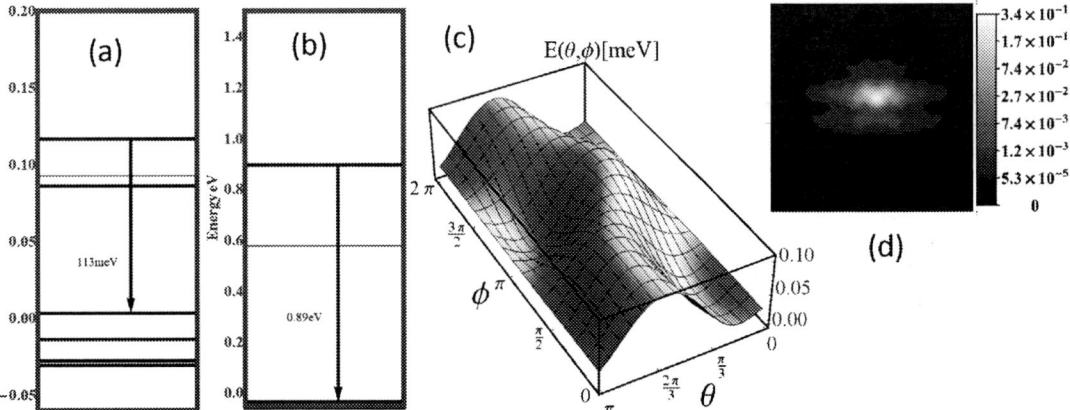

Figure 1 – Electronic properties of Mn in GaAs for the case in which the Mn d-levels are included explicitly. (a) Mn in bulk. The red line is the highest occupied level. The black line above that is the acceptor. (b) Mn on the (110) surface. (c) Magnetic anisotropy energy landscape. (d) The (110) cross-sectional view of the LDOS of the acceptor state.

In Fig. 1 we plot results for one Mn impurity in GaAs. The first panel refers to one Mn in bulk. After using *ab-initio* results for a first estimate of the onsite energies for Mn d-levels, we tune the T_2 values in such a way that the acceptor level (black line above the red line) occurs at the expected 113 meV above the valence band edge. Panel (b) of the same figure shows results for a Mn on the (110) surface, where the same bulk parameters are used. We find that, in agreement with previous results, the acceptor is a deep state positioned at about 0.85 eV above the valence-band edge, in remarkable agreement with STM experiments. Panel (d) shows the strongly anisotropic LDOS of the surface acceptor level, again in good agreement with the results of Ref. [12] and experiment. Finally, the magnetic anisotropy landscape is plotted in Fig. 1(c). (The coordinate system used for this plot has $\theta = 0$ parallel to the [001] axis and ($\theta = \pi/2$, $\phi = \pi/2$) parallel to [010].) In general we find that the magnetic anisotropy for Mn positioned in successive sublayers below the surface is in good agreement with previous results [12]. This implies that the model based on the classical spin description for the dopant spin works well for Mn in GaAs.

Fe dopants in GaAs

We now come to the discussion of Fe dopants in GaAs. In GaAs Fe ions are known to exist both in the $[Fe^{2+}]^-$ ($3d^6$) charged state, and in the $[Fe^{3+}]^0$ ($3d^5$) neutral isoelectronic state, depending on the Fermi level [15]. STM experiments [16] have investigated this system and recent work [17] shows that it is possible to manipulate the charge state of the dopant with the STM tip. The presence of an extra electron in the minority d-levels makes the situation more complicated than Mn and the classical spin model is much less justified. Therefore our analysis is based on Eq. (1) with Fe d-levels properly included. We first consider the case of bulk Fe. In Fig. 2, we plot the d-level density of states (DOS) at the Fe site, as obtained by DFT calculations using Wien2K [18]. In bulk calculations we consider a supercell made of 2x2x2 GaAs unit cells containing a total of 64 atoms. We find that the splitting between minority E levels and T_2 levels above the Fermi level (Energy = 0) is significantly affected by the Hubbard U parameter. In particular, for U=0 the degenerate E doublet is lower in energy that the T_2 triplet. The two levels swap when U = 4 eV. In the literature [15] the minority doublet E level is predicted to lie at 510 meV above the valence-band edge, while the minority triplet T_2 level is found to be a further 370 meV higher than the E.

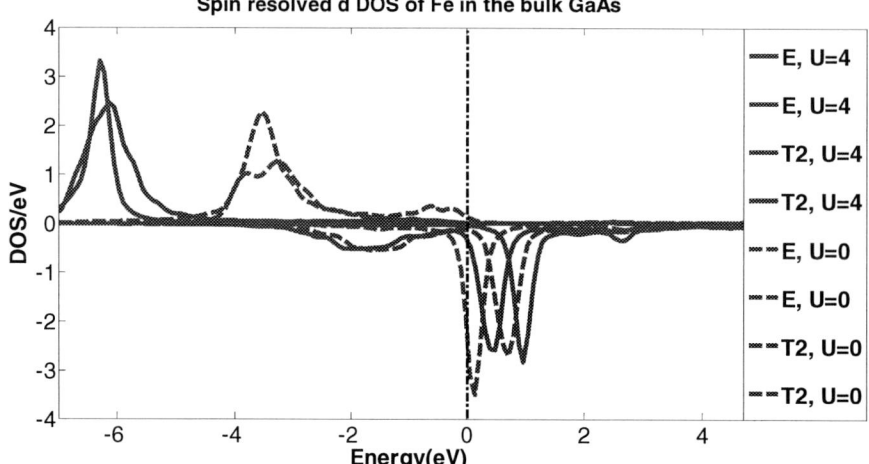

Figure 2 – DOS for a Fe atom in bulk GaAs, calculated by DFT using the Wien2K package [18]. Note that the relative position of the T_2 and E level above the Fermi level depends on the inclusion of a Hubbard U.

We have implemented this level structure in our tight-binding model by tuning mainly the splitting between majority and minority T_2 levels. The in-gap level structure is shown in Fig. 3(a). Note that expected level symmetry imposed by the tetragonal environment is perfectly reproduced by the tight-binding model. According to a simple electron counting in our finite cluster, all these five d-like levels in the gap are empty. The highest occupied level is quite close to the valence –band edge. Therefore the system is the $[Fe^{3+}]^0$ charge state. The (110) cross-sectional LDOS of the two E and the three T_2 levels is shown in Fig. 3(b). In some cases the LDOS strongly resembles the form of the LDOS of the Mn acceptor.

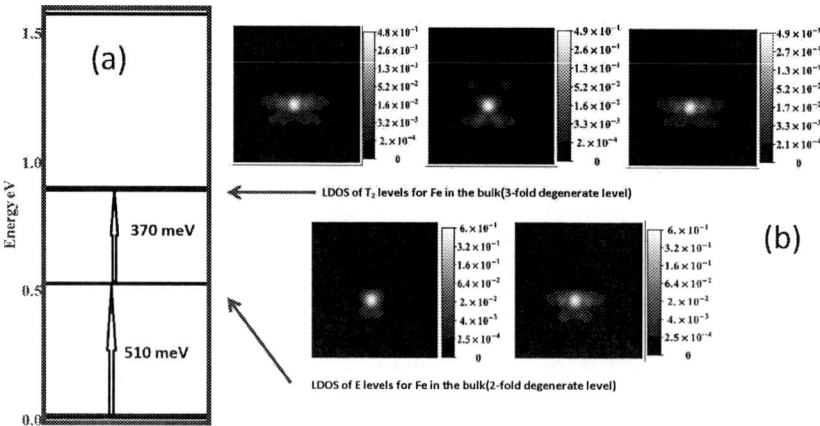

Figure 3 – (a) Electronic structure for a Fe in GaAs showing the three T_2 and the two E degenerate levels inside the gap. (b) (110) Cross-sectional LDOS for these five d-levels.

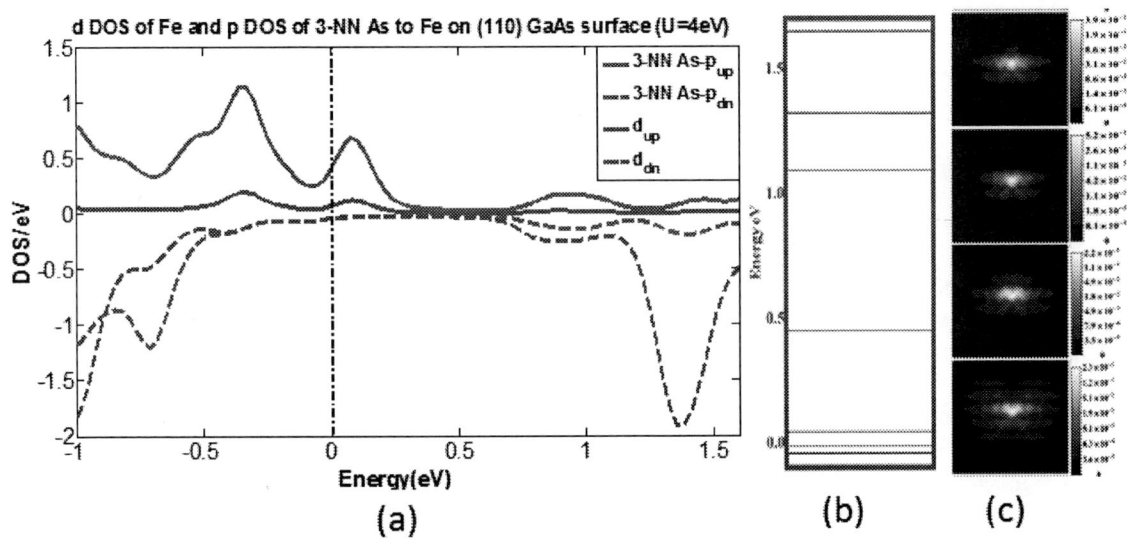

Figure 4 – (a) p- and d-resolved DOS for a Fe dopant on the (110) GaAs surface, calculated from *ab-initio* methods. (b) Ensuing electronic structure in the GaAs gap calculated from a tight-binding (TB) model. The TB parameters have been extracted from the *ab-initio* calculations. (c) (110) surface LDOS for some of the states inside the gap.

To study study Fe impurities on the (110) surface of GaAs we first use *ab-inito* calculations to obtain information on how the symmetry is broken, as well as input parameters for the TB model. The (110) surface of GaAs is constructed by cutting the bulk crystal along < 110 > direction. The surface supercell consists of six layers with 18 atoms at each layer, *i.e.* a total of 96 atoms. The supercell in the presence of the dopant is fully relaxed. A plot of the p- and d-resolved density of states at the As and Fe positions respectively is shown if Fig. 4(a). In these calculations we always include a Hubbard U = 4 eV, which is crucial to obtain the relative position of the spin polarized d-levels. In Fig. 4 (b) we plot the electronic structure of the energy levels in the gap as derived from the TB model calculations. In the gap we find five levels above the highest occupied state (red line), which are spatially localized around the impurity. The topmost level occurring at 1.6 eV is a delocalized level belonging to the conduction band. In Fig. 4(c) we plot the LDOS for four of these localized levels (top plot is for the level at ~1.3 eV). STM experiments [16] find three main resonances in the gap occurring at energies 0.85, 1.1 and 1.5 eV respectively. The energies levels in our calculations found at ~0.5, 1.0 and 1.3 eV are in qualitative agreement with this picture, albeit they occurr at slightly lower values. The shape of the LDOS for these three levels is also in qualitative agreement with experiment. The success of this approach indicates that the TB model developed here might be used to elucidate more complicated features recently unearthed in STM experiments on Fe dopants in GaAs [17].

SINGLE DOPANTS IN TOPOLOGICAL INSULATORS

In this section we present the results of ongoing efforts towards realistic TB modeling of electronic and magnetic properties of single dopants in three-dimensional (3D) topological insulators (TIs) [7,8], with a focus on Bi_2Se_3. 3D TIs are characterized by the non-trivial bulk insulating gap and topologically protected surface states with linear dispersion, traversing the gap [9,10]. An important question that has stimulated much theoretical and experimental research is how the topological surface states are modified in the presence of external perturbations [19]. Of particular interest is the case of time-reversal-breaking perturbations, such as magnetic impurities [20-24]. The purpose of this work is to provide a quantitative description of the interplay between surface states and individual dopants using a microscopic TB model. Below we present the first steps made in this direction.

Tight-binding model for Bi_2Se_3

The crystal structure of Bi_2Se_3 [25] can be most easily visualised as a periodic stacking of quintuple layers (QLs), formed by five consecutive layers of hexagonally arranged Bi and Se atoms [see inset in Fig. 5(a)]. The electronic structure of Bi_2Se_3 is described using the sp^3 TB model, similarly to the case of (Ga,Mn)As investigated in the previous section. We adopt the parameterization of the TB Hamiltonian obtained by Kobayashi [26] by fitting to DFT. We include interactions between atoms in the same atomic layer and between atoms in first and second nearest-neighbours layers, which is necessary to obtain a good fit to the DFT bulk band structure. The spin-orbit interaction is incorporated in the TB Hamiltonian in the same way as for GaAs [see Eq. (1)].

Surface band structures

Figure 5 shows the surface band structures of infinite slabs of Bi_2Se_3 with varying thickness calculated using the TB model. A finite gap is found for slab thicknesses below 5QLs. This is a recently discovered feature of TI thin films which has been attributed to valence-to-conduction band interactions between surface states localized on the opposite surfaces of the film [27]. The size of the gap decreases with increasing the film thickness. Such gap-opening mechanism has been considered for possible applications in TI-based MOSFET devices [28]. We obtained the following values of the gap: (a) $\Delta=0.8369$ eV for 1QL, (b) $\Delta=0.1788$ eV for 2QL, (c) $\Delta=0.0433$ eV for 3QL and (d) $\Delta=0.0074$ eV for 4QL. For slab thicknesses exceeding 5QLs, the gap is of the order of or smaller than 10^{-5} eV and for thick slabs [Fig. 5(f)] one recovers a typical surface band structure of Bi_2Se_3 with metallic surface states within the bulk band gap and a single Dirac point. Note that the surface band structures shown in Fig. 5 are identical to those obtained by Kobayashi [26] and are similar to other calculations [29], also bearing a close resemblance to experimental APRES data [10].

Finite-cluster approach

We further discuss the results of the TB modeling of Bi_2Se_3 thin films carried out using the finite cluster approach. Here an infinite slab with a given thickness is represented by a large but finite cluster with the same thickness. Such construction gives a reasonable approximation to the infinite slab and will allow us to investigate single impurities. A 1QL-thick rectangular cluster is shown in Fig. 6(c). The total size of the cluster for a given thickness can be controlled by changing the size in x and y directions, which lay in the (111) surface plane. Periodic boundary conditions are applied in the x-y plane in order to mimic an infinite structure and to eliminate any spurious energy states originating from the atoms on the edges. Hence the resulting geometry of the cluster is that of a torus with a finite thickness. The electronic structure is analyzed in the form of eigenvalue spectra, obtained by diagonalization of the TB Hamiltonian, which are then compared with the corresponding surface band structures. Finite cluster calculations have been carried out for slabs with thickness of 1-5 QLs and with the total number of atoms varying from 100 to 15000. A snapshot of these calculations is presented in Fig. 6.

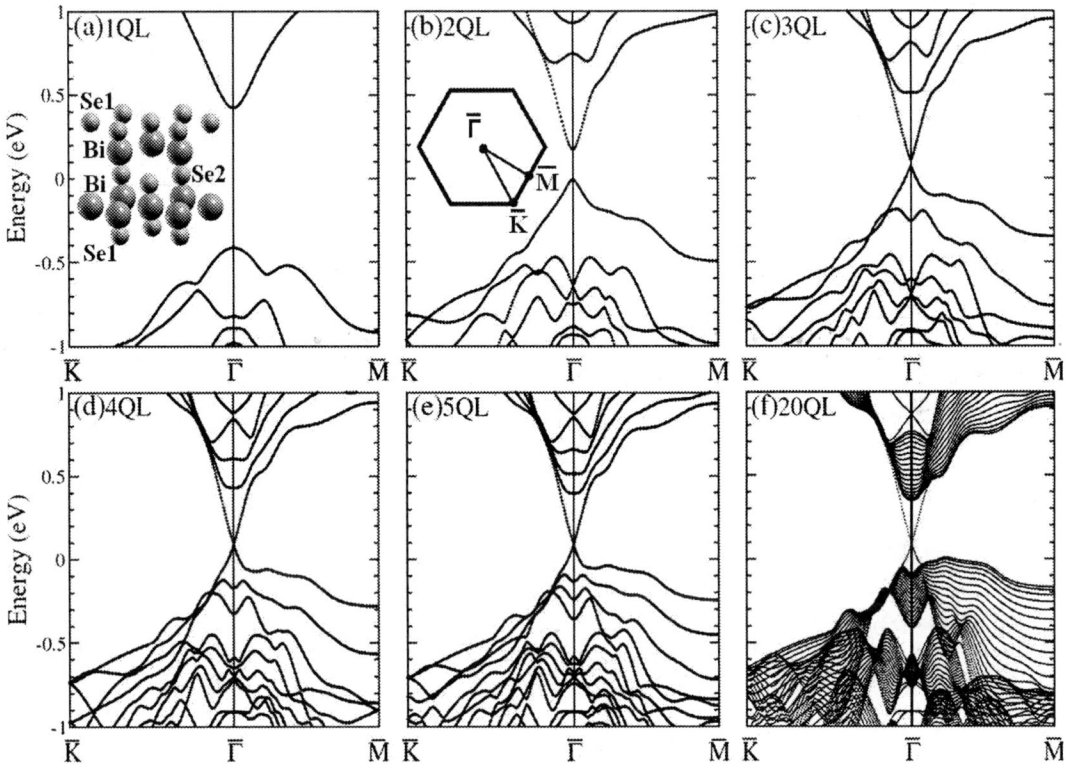

Figure 5 -- Band structures of a slab of Bi_2Se_3 of varying thickness: (a) 1QLs, (b) 2QLs, (c) 3QLs, (d) 4QLs, (e) 5QL and (f) 20QL obtained with the TB model. The inset of panel (a) shows the structure of 1QL of Bi_2Se_3. The inset of panel (b) shows the two-dimensional Brillouin zone of (111) surface of Bi_2Se_3.

The eigenvalue spectrum of a 1Q-thick cluster [see Fig. 6(a)] reveals a clearly defined energy gap. The size of the gap is in a good agreement with that obtained from the surface band structure [Fig. 5(a)]. Similar agreement between finite cluster and band structure calculations is also found for 2QL. More subtle is the case of thicker slabs ($\geq$ 3QL), for which clear regions of linear dispersion start two form in the surface band structure [see Figs. 5(b)-(d)] until the Dirac point finally emerges for 5QLs [Fig. 5(e)]. Due to the finite size of the system (and hence the discreteness of the energy spectrum), increasingly larger clusters are required to reproduce the linear dispersion. In the case of a 5QL slab, for instance, the maximum cluster size that we are currently able to tackle (~13000 atoms) is still not enough to reproduce the linear dispersion. However, by analyzing the eigenvalue spectra for increasing number of atoms we confirmed the trend towards linear dispersion: with increasing the size of the cluster more and more states start to fill the energy window corresponding to the linear dispersion. This is an interesting property of the finite cluster calculations that will be investigated further.

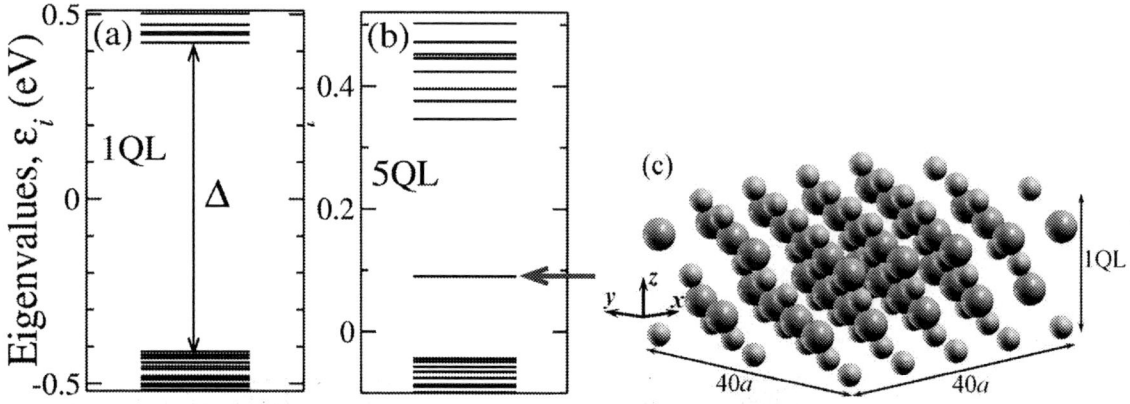

Figure 6 -- Eigenvalue spectra of a finite cluster of Bi_2Se_3 with 1QL (a) and 5QL (b) thickness. The size of the cluster in (a) is $40a$ x $40a$ in the (111) plane, which corresponds to the total of 8182 atoms. Here a is the hexagonal lattice constant. The size of the cluster in (b) is $22a$ x $22a$, which corresponds to the total of 12573 atoms. Note the energy gap Δ=0.8369 (eV) for 1QL and the position of the Dirac point at ~0.09 eV for 5QLs (red arrow). (c) A schematic of the 1QL-thick cluster used for calculation in (a).

Yet another remarkable feature is that despite the difficulties in reproducing the linear dispersion the nature of the states located exactly at the Dirac point is well captured by the finite cluster calculations. The eigenvalue spectrum of the 5QL slab, shown in Fig. 6(b), reveals distinct energy states in the bulk band gap whose position coincides, within the error of 10^{-5} eV, with the position of the Dirac point (~0.09 eV) in the corresponding surface band structure [see Fig. 5(e)]. Further analysis has confirmed that these states are two Kramers doublets localized on the opposite surfaces of the slab. This is a strong indication that the finite cluster TB model is able to capture some of the key properties of topological surface states in this system.

Bi_{Se} antisite defect

Finally, as an application of our finite cluster approach, we consider a 1QL-thick slab of Bi_2Se_3 with a Bi substituting for a Se atom (Bi_{Se} antisite defect) in the topmost atomic layer (see the inset in Fig. 7). In order to model the defect, the onsite energies of the Se atom are substituted by that of Bi. This particular setup is motivated by an experimental study by Urazhdin *et al.* [30], where STM spectroscopy was performed on the surface of Bi_2Se_3 in the presence of excess Bi. The topographic images revealed bright clover-shaped features which have been linked to Bi_{Se} antisite defects located in the fifth atomic layers below the surface.

Figure 7 shows the comparison between the eigenvalue spectrum of the pristine slab and that of a slab with the defect. In the latter case few states appear in the gap near the valence band edge. In order to investigate further the origin of these states we calculate atom-projected LDOS, integrated over the energy window containing these states, for all five atomic layers forming a 1QL slab (Fig. 8). Moving further away from the first monolayer, where the defect is located, the defect-induced spatial features in LDOS attain a characteristic clover-like shape, similar to topographic images of defects found in Ref. [30].

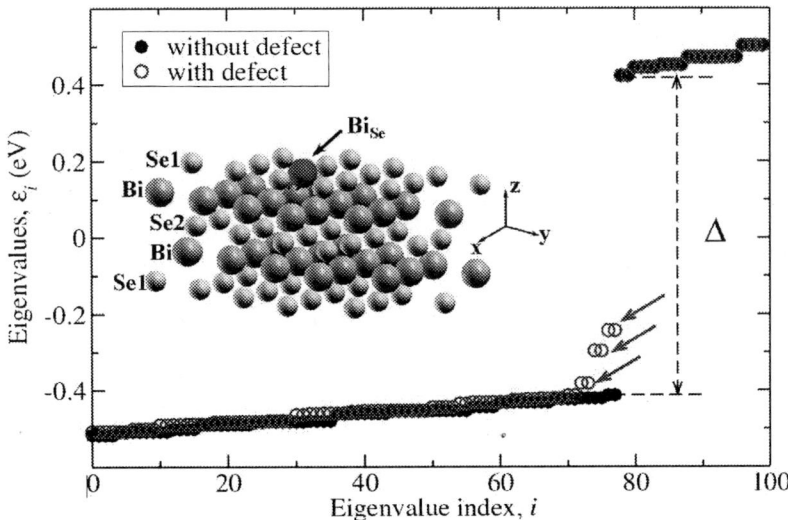

Figure 7 -- Eigenvalue spectrum of a finite cluster representing a 1QL-thick slab of Bi_2Se_3 with and without the defect. In the case of the pristine slab the energy gap Δ is indicated in the figure. In the case of the slab with a defect red arrows mark the eigenvalues appearing in the gap. The inset shows a part of the cluster with a total of 8182 atoms, used in the calculation. Note the position of the Bi_{Se} antisite defect in the first Se monolayer.

A particular shape of LDOS maps for atomic layers 2 to 5 [Fig. 8(b)-(e)] can be explained based on a simple model of the chemical bonding within QL structure [25]. Each of the three p-orbitals of an atom points towards one of the three nearest-neighbor atoms in the layer above and in the layer below, forming almost regular octahedral environment. Hence the overlapping p-orbitals of atoms in adjacent atomic layers form strongly interacting $pp\sigma$ chains [Fig. 8(f)]. The defect introduces a disturbance along the three $pp\sigma$ directions. Away from the defect, this disturbance can be seen through the atoms terminating three $pp\sigma$ chains emanating from the defect atom, producing a clover-like feature in the LDOS.

CONCLUSIONS

In this paper we have investigated the electronic structure of individual dopants placed on the surface of two important quantum materials, which is accessible to STM experiments.

In the first example we have considered TM impurities in GaAs. A tight-biding model built with the help of *ab-initio* calculations is able to capture some of the important features of the midgap states associated with the magnetic impurities.

In the second example we have considered as a host material $Bi_2Se_3\theta$, which is an archetypal three-dimensional topological insulator. We have employed the sp^3 tight-binding model for Bi_2Se_3 with parameters extracted from DFT to calculate the electronic structure of infinite slabs and finite clusters with impurities. The calculated thickness-dependence of the surface band structures is in good agreement with results available in the literature. As a further aspect, we have investigated the effect of a substitutional Bi defect on the electronic structure of a 1QL thin film of Bi_2Se_3. We find similarities between the calculated LDOS and topographic images observed in STM experiments, which can be explained using a simple model of chemical bonding in Bi_2Se_3.

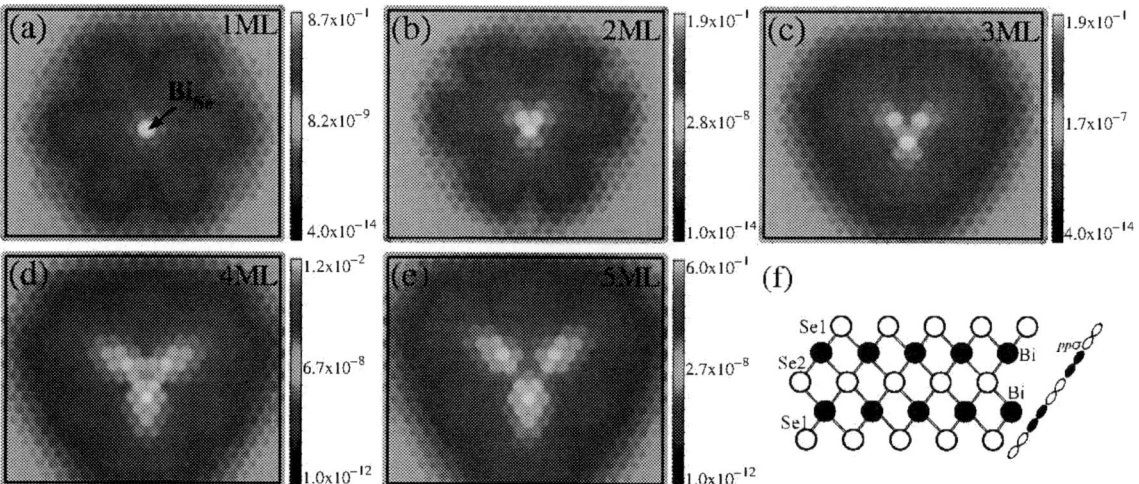

Figure 8 -- LDOS plotted for five monolayers (ML) forming a 1QL-thick slab of Bi_2Se_3 with 8182 atoms and with a Bi_{Se} antisite defect located in the first ML [panels (a)-(e) correspond to ML 1-5]. The LDOS is integrated over an energy window [-0.4 eV;-0.2 eV] (see Figure 3). The position of the defect is indicated in panel (a). The images cover the area with dimensions $12a$ x $12a$, with the defect located in the center.

REFERENCES

1. P. M. Koenraad and M.E. Flatté, *Nat. Mater.* **10**, 91 (2011).
2. D.D. Awschalom, N. Samarth and D. Loss, eds., *Semiconductor Spintronics and quantum computation*, Springer-Verlag, Berlin (2002).
3. D.D. Awschalom and M.E. Flatté, *Nat. Phys.* **3**, 153 (2007).
4. D. Loss and D.P. DiVincenzo *Phys. Rev. A* **57**, 120 (1998).
5. A. M. Yakunin *et al., Phys. Rev. Lett.* **92**, 216806 (2004).
6. D. Kitchen *et al., Nature* **442**, 436 (2006).
7. M. Z. Hasan and C. L. Kane, *Rev. Mod. Phys.* **82**, 3045 (2010).
8. X.-L. Qi and S.-C. Zhang, *Rev. Mod. Phys.* **83**, 1057 (2011).
9. H. Zhang *et al., Nature Physics* 5, 438 (2009).
10. Y. Xia *et al, Nature Physics* **5**, 398 (2009).
11. J.-M. Tang and M.E. Flatté, *Phys. Rev. B* **72**, 161315 (2005).
12. T.O. Strandberg, C.M. Canali and A.H. MacDonald., *Phys. Rev. B* **80**, 024425 (2009).
13. M. F. Islam and C.M. Canali, *Phys. Rev. B* **85**, 155306 (2012).
14. J.-M. Jancu *et al., Phys. Rev. B* **57**, 6493 (1998).
15. E. Malguth *et al., Phys. Status Solidi B* **245**, 455 (2008).
16. A. Richardella *et al., Phys. Rev. B* **80**, 045318 (2009).
17. J. Bocquel *et al., Phys. Rev. B* **87** 075421 (2013).
18. P. Blaha, K. Schwarz, G. K. H. Madsen, D. Kvasnicka, and J. Luitz, WIEN2k, An Augmented Plane Wave Plus Local Orbitals Program for Calculating Crystal properties (Vienna University of Technology, Austria) (2001).
19. L. A. Wray *et al., Nature Physics* **7**, 21 (2011).
20. Q. Liu, C. X. Liu, C. Xu, X. L. Qi, and S.-C. Zhang, *Phys. Rev. Lett.* ***102***, 156603 (2009).
21. H. Beidenkopf *et al., Nature Physics* **7**, 939 (2011).
22. D. Zhang *et al, Phys. Rev. B* **86**, 205127 (2012).

23. J.-M Zhang, W. Zhu, Y. Zhang, D. Xiao, and Y. Yao, *Phys. Rev. Lett.* **109**, 266405 (2012).
24. Y. S. Hor *et al.*, *Phys. Rev. B* **81**, 195203 (2010).
25. S. K. Misra, S. Satpathy, and O. Jepsen, *J. Phys.: Condens. Matter* **9**, 461 (1997).
26. K. Kobayashi, *Phys. Rev. B* **84**, 205424 (2011).
27. Y. Zhang *et al.*, *Nature Physics* **6**, 584 (2010).
28. J. Chang, L. F. Register, and S. K.Banerjee, *J. Appl. Phys.* **112**, 124511 (2012).
29. W. Zhang, R. Yu, H.-J. Zhang, X. Dai, and Z. Fang, *New. J. Phys.* **12**, 065013 (2010).
30. S. Urazhdin, D. Bilc, S.H. Tessmer, and S. D. Mahanti, *Phys. Rev. B* **66,** 161306(R) (2002).

AUTHOR INDEX

Asenov, Asen.. 1

Canali, C. M. 13

Georgiev, Vihar.. 1

Islam, M. F.. 13

MacDonald, A. H. 13

Mahani, M. R. 13

Pertsova, A... 13